Name: Eng. / Mostafa Yacoub
Abdellatif Mahmoud
Nationality: Egyptian
ORCID: 0000-0002-9991-4624
Qualification: civil engineer Cairo
University 2003

I0477897

- **<u>Reciprocal of positive integers:</u>**

  In this paper, we will discover the relation between the summation of the reciprocal of positive integers and the summation of positive integers that depends on prime numbers based on my discovered formula that connects prime numbers to composite numbers that belong to the Array of odd numbers (Array PTBP) within ranges that extend to infinity.

- ## My discovered formula:
Definitions:
Array PTBP

It is the following Array of odd numbers

$$\begin{vmatrix} 1 & 3 & 7 & 9 \\ 11 & 13 & 17 & 19 \\ 21 & 23 & 27 & 29 \\ 31 & 33 & 37 & 39 \\ 41 & 43 & 47 & 49 \\ 51 & 53 & 57 & 59 \end{vmatrix}$$

And so on....

- **For a given set of consecutive primes whose numbers =n that start with prime 3 and end with prime F and not including prime 2 and prime 5**
i.e.
set=[3,7,11,13,..
........................,F]
S=product of those consecutive primes
i.e

$$S = \prod_{i=3}^{i=F} (i)$$

Range=R_k = 10 × S × k

Where k=[1, 2, 3, 4, ..., ∞(infinity)

i.e R_1=10 x S x 1 and R_2=10 x S x 2

And so on

- Number of composite numbers that belong to Array PTBP and created by the effect of those consecutive primes within the range R_K

- $=[(K \times 4^{\times \frac{S}{3}}) + (\sum_{j=7}^{j=F} (K \times 4 \times (\frac{S}{j}) \times \prod_{i=7}^{i = \text{prime number before current prime number } j} (\frac{i-1}{i})$
)]-(n)

Where j =consecutive values of primes

7, 11, 13,..............., F

And i= consecutive values of primes 3, 7, 11, 13,…….., prime before current j prime

The previous formula can be applied for any number of consecutive prime numbers that start with prime number 3

- The first term $(k \times 4 \times \frac{S}{3})$ represents the count of unique Composite numbers +1 that belong to the Array PTBP and are created by prime number 3 within the range

$$R_k = 10 \times S \times k$$

- **The second term**

$$\sum_{j=7}^{j=F} (K \times 4 \times (\frac{S}{j}) \times \overset{i = \text{prime number before current prime number } j}{\prod_{i=7}} (\frac{i-1}{i})$$

Represent the count of unique Composite numbers+n-1 that belong to the Array PTBP and are created by each prime number after the prime number 3 within the range

$R_k = 10 \times S \times k$

- **The third term (-n)**
Subtracting n (number of consecutive primes starting from prime number 3) because the count of composite numbers generated from those consecutive primes includes the count of those primes in the range

$R_k = 10 \times S \times k$

- Explanation and proof for my theory in my previous paper (prime number theory)

- We will mention only the concept of the number cycle

  We can use the number cycle concept to understand the behavior of consecutive primes in creating composite numbers.

  i.e.

  $$S = \prod_{i=3}^{i=F} (i)$$

  Range=cycle range= R_k = 10 × S × k

  Where k= [1, 2, 3, 4, ...,∞(infinity)

  i.e. R_1=10 x S x 1 and R_2=10 x S x 2

  And so on

- Now consider only one k value =1
- And Now
- For any set of consecutive primes

  i.e.

  set=[3,7,11,13,.............................

  ....,F]

S=product of those consecutive primes

i.e

$$S = \prod_{i=3}^{i=F} (i)$$

Range=R_k = $10 \times S \times k$

- Any cycle containing two types of numbers

  First type (numbers that are divisible by prime numbers within the set).

  The second type (numbers that are not divisible by prime numbers within the set).

- For example, the following figure (considering the set of consecutive primes=[3,7]) shows the two types of numbers the colored one represents the first type while the uncolored numbers represent the second type

- We can see there must be a repeated pattern for each type of number and for both of them together each cycle up to infinity i.e for

  k= [1, 2, 3, 4, ……………., ∞(infinity)

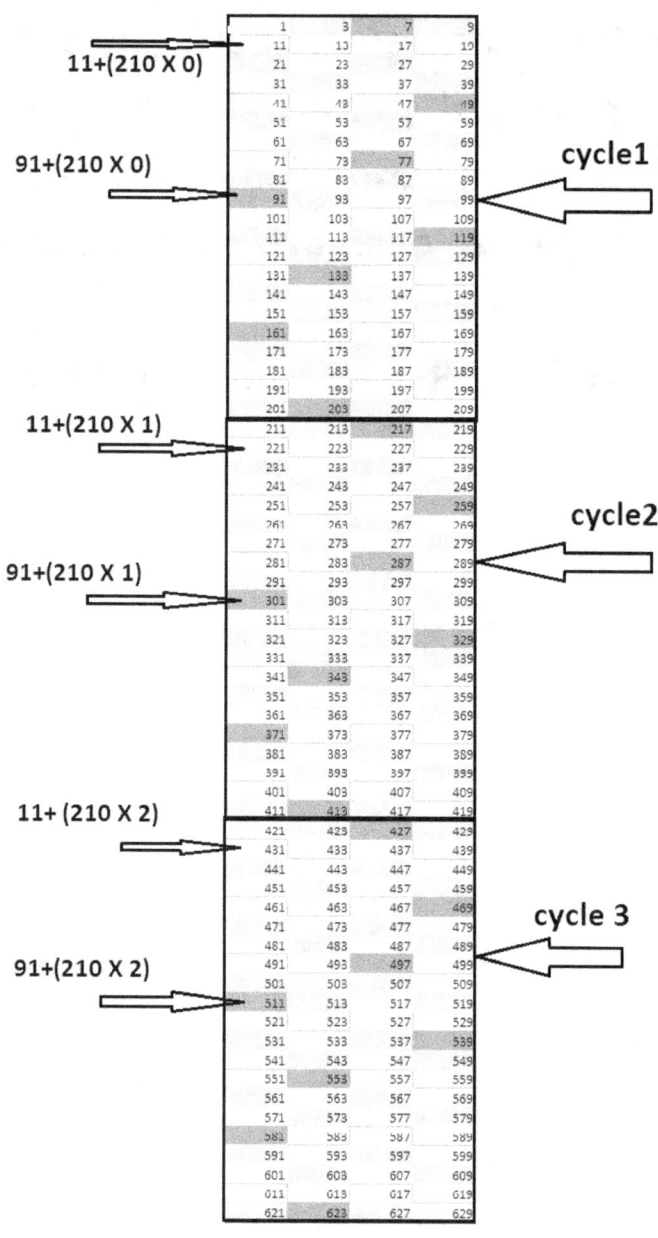

- **We have proved (in my paper prime number theory part3) that the Summation of all numbers within Range = S x 10**
Including all numbers of the array PTBP (with last digit= 1 or 3 or 7 or 9) and all other even or odd numbers that are divisible by two or five (with last digit = 0 or 2 or 4 or 6 or 8 or 5)

$$=[[\ 50\ x\ (\prod_{i=3}^{i=F}(i\ x\ i)\ )\ ]+[5\ x\prod_{i=3}^{i=F}(i)\ ]]$$

- And for different values of K =[1,2,3,............, ∞(infinity)

$$=[[\ 50\ ^x\ (\prod_{i=3}^{i=F}(ixi)\ )\ ]\ ^x\ [K^2]]$$

$$+[5\ ^{x\ K\ x}\prod_{i=3}^{i=F}(i)\ ]$$

- Example 1

  For the set of consecutive prime numbers = [3]

  The summation of all numbers that belong to the following array within the range

  = 10 x 3 = 30

1	2	3	4	5	6	7	8	9	10
11	12	13	14	15	16	17	18	19	20
21	22	23	24	25	26	27	28	29	30

$$=[50 \times 3 \times 3] + [5 \times 3] = 450 + 15 = 465$$

- **Example 2**

For the set of consecutive prime numbers = [3] and for two cycles i.e k=2

The summation of all numbers that belong to the following array within the range

= 2 x 10 x 3 = 60

1	2	3	4	5	6	7	8	9	10
11	12	13	14	15	16	17	18	19	20
21	22	23	24	25	26	27	28	29	30
31	32	33	34	35	36	37	38	39	40
41	42	43	44	45	46	47	48	49	50
51	52	53	54	55	56	57	58	59	60

$$=[50 \times 3 \times 3 \times (2^2)] + [5 \times 3 \times 2] = 1800 + 30$$
$$= 1830$$

- **Example 3**

  For the set of consecutive prime numbers = [3] and for three cycles i.e k=3

  The summation of all numbers that belong to the following array within the range

  $$= 3 \times 10 \times 3 = 90$$

1	2	3	4	5	6	7	8	9	10
11	12	13	14	15	16	17	18	19	20
21	22	23	24	25	26	27	28	29	30
31	32	33	34	35	36	37	38	39	40
41	42	43	44	45	46	47	48	49	50
51	52	53	54	55	56	57	58	59	60
61	62	63	64	65	66	67	68	69	70
71	72	73	74	75	76	77	78	79	80
81	82	83	84	85	86	87	88	89	90

$$=[50 \times 3 \times 3 \times (3^2)] + [5 \times 3 \times 3] = 4050 + 45$$

$$= 4095$$

- **Example 4**

  For the set of consecutive prime numbers = [3,7]   and for one cycle i.e   k=1

  The summation of all numbers that belong to the following array within the range

  = 1 x 10 x 3 x 7 = 210

1	2	3	4	5	6	7	8	9	10
11	12	13	14	15	16	17	18	19	20
21	22	23	24	25	26	27	28	29	30
31	32	33	34	35	36	37	38	39	40
41	42	43	44	45	46	47	48	49	50
51	52	53	54	55	56	57	58	59	60
61	62	63	64	65	66	67	68	69	70
71	72	73	74	75	76	77	78	79	80
81	82	83	84	85	86	87	88	89	90
91	92	93	94	59	96	97	89	99	100
101	102	103	104	105	106	107	108	109	110
111	112	113	114	115	116	117	118	119	120
121	122	123	124	125	126	127	128	129	130
131	132	133	134	135	136	137	138	139	140
141	142	143	144	145	146	147	148	149	150
151	152	153	154	155	156	157	158	159	160
161	162	163	164	165	166	167	168	169	170
171	172	173	174	175	176	177	178	179	180
181	182	183	184	185	186	187	188	189	190
191	192	193	194	195	196	197	198	199	200
201	202	203	204	205	206	207	208	209	210

$$=[50 \times 3 \times 3 \times 7 \times 7)] + [5 \times 3 \times 7] = 22050 + 105$$

$$= 22155$$

- **Example 5**

  For the set of consecutive prime numbers = [3,7] and for one cycle i.e k=3

  The summation of all numbers that belong to the following array within the range

  = 3 x 10 x 3 x 7 = 630

=[50 x 3 x 3 x 7 x 7 x(3^2))] + [5 x 3 x 7 x 3] = 198450 + 315

 = 198765

- **And now we have**

$$\sum_{n=1}^{\infty}\left(\frac{1}{(n^2)}\right) = \left(1 - \frac{1}{(2^2)}\right) \text{ x } \left(1 - \frac{1}{(3^2)}\right) \text{x}$$

$$\left(1 - \frac{1}{(5^2)}\right)$$

$$\text{x}\left(1 - \frac{1}{(7^2)}\right) \text{x} \left(1 - \frac{1}{(11^2)}\right) \text{x} \left(1 - \frac{1}{(13^2)}\right)$$

x ....

$$= (\pi^2)/6 = \prod_{p\,prime} \left(1 - \frac{1}{(P^2)}\right)$$

Where p takes values 2,3,5,7,11,13,......

And

$$A = \left(1 - \frac{1}{(2^2)}\right) \text{ x } \left(1 - \frac{1}{(5^2)}\right)$$

$$x \prod_{p\,prime} \left(1 - \frac{1}{(P^2)}\right)$$

$$A = \left(\frac{18}{25}\right) \prod_{prime} \left(1 - \frac{1}{(P^2)}\right)$$

Where p takes values that belong to the array PTBP

i.e p = 3,7,11,13,17,.......

And now

- For a given set of consecutive primes whose numbers =n that start with prime 3 and end with prime F and not including prime 2 and prime 5

  i.e.

  set=[3,7,11,13,..

  ....,F]

  we have

The summation of all numbers within Range = S x 10

Including all numbers of the array PTBP (with last digit= 1 or 3 or 7 or 9) and all other even or odd numbers that are divisible by two or five (with last digit = 0 or 2 or 4 or 6 or 8 or 5)

$$B=[[\ 50 \ x \ (\prod_{i=3}^{i=F}(ixi) \ ) \ ] \ x$$

$$[K\char`\^2]]$$

$$x \ K \ x \prod_{i=3}^{i=F}(i)$$

$$+ [5 \qquad]$$

Where K =[1,2,3,............,
∞(infinity)

we can put K = $\left[\displaystyle\prod_{p\ prime} (P) \right] / \left[\displaystyle\prod_{i=3}^{i=F} (i) \right]$

And we have

$$\frac{\sum_{n=1}^{\infty}\left(\frac{1}{(n^2)}\right)}{\sum_{n=1}^{\infty}(n)} = \frac{A}{B}$$

$$= \left(\frac{18}{25}\right) \; x \prod_{p\,prime}\left(1 - \frac{1}{(P^2)}\right) \Big/$$

$$[[50 \; x \; (\prod_{i=3}^{i=F}(i\char`\^2\,))] \; x \; [K\char`\^2]]$$

$$+ [5 \quad x\,K\,x \prod_{i=3}^{i=F}(i) \quad]$$

Where $K = \dfrac{\left[\displaystyle\prod_{p\,prime}(P)\right]}{\left[\displaystyle\prod_{i=3}^{i=F}(i)\right]}$

And where both p and i take values that belong to the array PTBP

i.e P or i = 3,7,11,13,17, ...

www.ingramcontent.com/pod-product-compliance
Lightning Source LLC
Chambersburg PA
CBHW030046230526
45472CB00005B/1701